THE ATTRACTION FACTOR IN EXECUTIVE SUCCESS

Relationships in your office

Copyright by National Institute of Business Management, Inc.
1328 Broadway, New York, NY 10001

Introduction

Welcome to the 1990s, where the executive office is no longer an exclusive men's club. Women are being admitted in growing numbers. There are new rules. Situations unthinkable in the past now occur on a daily basis:

Man vs. Woman: The battle of the sexes takes on new meaning as men and women compete for advancement. But old prejudices die hard. Can you change these attitudes when they impede your progress?

Woman vs. Woman: Not all women are created equal in the corporation's eyes. Some have had to fight for all they have, while others seem to inherit more than their fair share. How do you bring the two camps together?

Man and Woman: Sexual attractions are inevitable. How do you deal with them?

This Special Book will help you evaluate your acumen for managing yourself and other people—men and women—in this fast-changing workplace. With some thought given beforehand, you will be able to deal skillfully with any problem tossed your way. The end result will be enhanced relationships with your superiors, peers and subordinates. And that will spell advancement for you!

Table of Contents

Table of Contents (Cont'd)

Author	**Charlene Canape**
Manager, Desktop Publishing Operations	**Marie Mularczyk**
Editorial Assistant	**Brian T. Ehrlich**
Art Director	**John Kwong**
Graphics Coordinator	**Patricia Spieler**
Production Manager	**Edmund Leisten**
Production Assistant	**James Hipkiss**
Publisher	**Brian W. Smith**

Relationships In Your Office

Building solid relationships in today's workplace is a challenge. Years ago, bonds were formed over the water cooler while discussing last night's basketball game. A disparaging remark about a female employee's appearance may have received laughs all around. Secretaries were women; executives were men. The lines were clearly drawn and everyone understood the rules.

The flow of women into managerial ranks has changed all that. What may have been an innocuous remark between two workers in the past might now offend someone. An innocent pat on the back no longer seems so innocent.

You may have already found yourself smack in the middle of a gender-related dilemma. Consider the following:

Example Number One: A female vice president of marketing attends a meeting with four of her male counterparts and the senior vice president for the division, also a man. During the discussion, someone wonders aloud how a "housewife" would react to a specific advertising campaign. Heads turn and all eyes are trained on the woman executive. How does she react?

Example Number Two: A sales manager (male) visits a prospect with a newly recruited salesperson (female). Introductions are made and everyone sits down. The salesperson considers taking out her pad and pencil to take notes. She hesitates. Would such an action create the perception that she is a secretary?

Example Number Three: The head of personnel (male) for a major corporation arranges to host a Christmas lunch at a nearby club for his staff, which is all male except for one recently hired woman. His party arrives and, too late, he realizes that women are not allowed in the main dining room. Does he insist his party be seated, settle for second-class accommodations in another dining room, or suggest a nearby restaurant?

Example Number Four: A female executive accompanies her boss and several customers to lunch. While leaving the restaurant, her boss puts his arm around her and announces proudly, "Susan is the best assistant I've ever had," and then gives her a visible hug. Susan feels very uncomfortable, but doesn't want to embarrass her boss in front of the group. What should she do?

Example Number Five: At a manager's meeting someone tells a sexist joke. The joke teller is an insecure man who seems to have it in for female executives. His target at this meeting is a fast-rising woman he hopes to throw off guard. When she gets angry, he will be able to prove that she is uptight with no sense of humor. How does the woman manager react? More importantly, perhaps, how do the other men present react?

Being able to wend your way through one of these workplace thickets without getting pricked may decide your future. Will you demonstrate your superior abilities and move ahead?

This discussion is divided into two parts: Your Working Style and Your Working Relationships. Although the two areas are treated separately, they are connected. Your working style—how you present yourself in your office—dictates how others respond to you initially. Ultimately, however, your performance will be judged on more substantive issues, including how well you manage yourself and others in tricky situations.

Section One
YOUR WORKING STYLE

Your working style is a combination of many factors—your personality, your education, your experience, and your feelings about people. Too often managers get promoted and forget the very attributes that got them there. They try to emulate the attitudes, interests, and outlooks of all their colleagues. (Women, in particular, avoid topics that will label them as outsiders.)

Trying to conform to corporate culture is a mistake. If your business is inhospitable to your style and thinking, forcing yourself to adjust will only make you more uncomfortable.

Don't fall into that trap. Be yourself. If it doesn't feel comfortable, don't do it. Develop a management style that is yours and yours alone. Remember the qualities, characteristics, and traits—male and female—that could help you to become even better at your job.

Take the woman in Example One. While she should discourage her colleagues from describing her as a "housewife," it is more than likely that she would be able to accurately predict how a woman would perceive a certain ad. That information would benefit her and her company, and she should use it.

Here are some things to remember about your own personal situation:

- **Individuality.** Don't try to anticipate what other people might like—simply espouse enthusiastically what you *do*. While sharing pursuits with others is important, having an involved, enthusiastic life is what will make you attractive to others—not whether they share all your interests and pursuits.

- **Standards.** Carefully define what your business and personal principles really are—then stick by them staunchly, no matter how difficult the circumstances. If you were a man sitting in on the meeting in Example Five when a co-worker told a sexist joke, your sensibilities would have been as offended as the woman manager who was the target, and you would have earned her respect (and probably everyone else's) by speaking out against the remark.

- **Focus.** The ability to concentrate completely on a task and get it done is central to advancement. And unlike intelligence, this ability can be developed. Focus is also the most important part of what is often called "executive presence." To see it in action, observe the person of power in a meeting. While others are veering off toward other issues, this person will not be dissuaded from his or her goals. Think for a moment about the woman manager in Example Five. If she allows the rude joke to distract her, she will leave the meeting without accomplishing her purpose.

- **Appropriateness.** Treat serious issues with a serious attitude. While telling a few jokes to blow off steam may break the tension of difficult decisions, it will only make you appear uncommitted and—in the case of our joke teller above—insensitive and vindictive.

• **Sobriety.** Nothing will undermine your credibility more quickly than having a few drinks at lunch and returning to the workplace smelling of alcohol. *Even worse:* drinking and acting out of character at a holiday party or picnic. What you do at home is your affair—but what you do at work is your *business.* By the same token, inappropriate romantic or sexual liaisons will push you right off the corporate ladder (more on socializing and drinking later in the Book).

How You Appear to Others

First impressions are based on the superficial—clothes, hairdos, jewelry, posture. Your appearance makes a statement and sends out signals. Your office—what's on the walls, on your desk, how your furniture is arranged—says a lot to those who enter. Your voice and mannerisms can portray a person who is filled with self-confidence or held back by self-doubt. This is why some workers—male or female—will always be treated as second-class citizens while others will be accorded respect.

Making a good first impression is particularly important in today's business environment where you may only get one shot to impress someone. You may be attending a conference to make future contacts. You want each person you talk with to come away with a positive image of you and your capabilities. Chances are that what you say in the few minutes after you meet will be less important than how you look and act. Your clothes, posture, mannerisms and speech can help you make sure your card winds up in the Rolodex, and not the trash can.

Personal Appearance

Your aim is to present a businesslike appearance that others will find attractive but not seductive. For obvious reasons, clothing or accessories that might be construed as a sexual come-on are out. This includes elaborate hairdos, excessive cologne or perfume, tight clothing, low-cut blouses or dresses for women, and flashy jewelry. Your goal should be to call attention to your talents and accomplishments, not your attire.

If you doubt the impact clothing and accessories may have, let's consider for a moment the woman executive in Example One who is being singled out as the "housewife" by her male counterparts. Imagine that she is sitting in that meeting wearing a colorful, loose-fitting dress and sandals. Now imagine she is wearing a more conservative dress with a jacket and black, low-heeled pumps. In which instance will she more readily be able to distance herself from the notion that she is a "housewife"?

Presenting the best image is sometimes difficult in the current working environment. You want to appear attractive to others without crossing over the line and becoming so attractive that you invite a pass from the opposite sex or jealousy and envy from members of your own sex. Here are some pointers for managing that feat:

- **Come clean.** Dandruff, scuffed or worn shoes, five o'clock shadow, runs in hosiery—all are turnoffs.
- **Observe local customs.** What's "in" for New Yorkers may be considered outlandish in the Midwest or South. If you have doubts, opt for the more conservative route. It's better to be underdressed than to stand out.
- **Less is more.** If you are a woman, avoid carrying a large bulky briefcase and a large bulky purse. Streamline your appearance by carrying a briefcase that doubles as a purse.
- **Use props.** Buy an expensive pen and agenda book. When you use them to take notes (as the sales trainee in Example Two was considering doing) you will convey the notion that you are an executive, not a secretary.
- **Opt for subtlety.** A famous leather maker's advertisements read: "When your own initials are enough." You don't need to flaunt your designer watches, ties, or wallets to tell people how important you are. Low key is better.
- **Stand tall.** Good posture and a healthy stride do more than help you breathe better; they say something about you. The image of an executive who walks into a room at a good clip, with head erect and shoulders back, sets the stage for everything else he or she will do to boost both self-image and the image projected to others. Once the stage is set, it's far easier for others to expect and accept the poise and confidence that follows.
- **Accept your limits.** Today's leading men and women are not always the best looking. Instead, they project a personality, a sense of themselves that transcends their physical size and features. You can do the same in a

business setting.

On the other hand, if you are blessed with good looks, count them as an asset that can help you look the part for a leadership role. Also note that while good looks will undoubtedly earn you the attention of the opposite sex, studies show that attractive people are also viewed as charismatic by members of their own sex.

How to Decorate Your Office

Your office should function well for you. It should look clean and efficient, without piles of paper or other clutter around. But you must also pay close attention to your office's atmosphere. Does it convey an air of seriousness? Does it look like a place where some valuable thinking is taking place?

To decorate your office like a leader, obey the following rules:

- **Express your personality.** Your office should reflect your interests and pursuits. Include things that appeal to you. Looking for items to please other people only waters down the personality you're trying to convey, and can make you seem wishywashy. Never be ashamed of your tastes. If you love your classic 1967 Corvette, put a picture of it on the wall—even if your colleagues all have reproductions of Impressionist paintings.

- **Include things you understand.** If you put up a painting or a framed quote you like and know about, you'll create a positive impression when you explain it to people.

If you put up something you're supposed to like but really don't, you'll seem pretentious.

• **Avoid the commonplace.** Don't decorate your space with two or three innocuous, framed posters the mail room brings you—it's another way to ensure anonymity. By the same token, avoid cute cartoons at all costs. Displaying a cartoon of a monkey saying, "I Hate Mondays!" will only make people view you as a monkey who hates work.

• **Avoid creating an environment that is too masculine or too feminine.** If you are a female executive, steer away from decor that is too flowery. It will make your male colleagues uncomfortable and convey the notion that you are soft and romantic—okay for your living room but not for your workspace. Similarly, men should avoid the all-leather look reminiscent of furnishings in an exclusive men's club. Some soft touches—a brass lamp, a potted plant, chairs upholstered in striped fabric—will make the office inviting to all your co-workers.

If you have children, it's fine to display photos. Women, however, should be aware of how their supervisor will interpret too many pictures—as a large exhibition could serve as a reminder of family commitments. Men, on the other hand, will enhance their reputations for caring about people, the more photos they have on their desk. (In both cases, a photo that includes a spouse can serve to deter someone who has shown more than a business interest in you.)

• **Carefully arrange seating.** Some executives in the advertising industry, which is a collaborative business that depends on creativity for its survival, prefer to work at a desk that is a round table. When someone visits them, it is most often for a brainstorming session, and this circular arrangement says that all participants are equal and all their contributions are considered valuable.

Give some thought to your own seating. And don't discount how you want to be viewed by colleagues and subordinates vis à vis your sex. A woman executive who feels she needs to remind male subordinates of her rank might choose a high back chair for her desk chair and lower chairs and a couch for visitors. In this manner, she will always be higher than anyone seated in her office, thus better conveying her authority.

Business Etiquette—The Greeting

The issue first surfaced during the 1984 Presidential campaign when Walter Mondale and Geraldine Ferraro shared the Democratic ticket. Was it considered proper protocol for the Presidential candidate to kiss his Vice Presidential running mate?

Soon, the issue received more widespread coverage (*The Wall Street Journal* devoted a whole front page column to the questions raised) as businesspeople confessed that they were confronted with the same problem on a daily basis. You are at a convention with your boss and meet a former colleague of the opposite sex that you have not seen for several years. You extend your hand and soon find yourself in a warm embrace being kissed on the cheek. How do you react?

Or, you are the one with the urge to do the kissing. Do you hold back?

Unless the person you are greeting also happens to be your cousin from Duluth, it is best to be conservative and just shake hands. On the other hand, if you find yourself on the receiving end of a colleague's affection, it is best not to overreact and make the situation worse.

In general, these rules apply:

• **Refrain from touching colleagues of the opposite sex.** Don't put an arm around them, place your hand on their shoulder, or touch them in any other way, no matter how close you might feel to them. (Recall how Susan in Example Four above was made to feel uncomfortable by the physical attentions of her boss).

• **Don't kiss someone you don't know well.** The person's reaction may be surprise, embarrassment, or even resistance. Generally, the longer you've known a person and the more established your social and business relationships, the more appropriate a friendly peck on the cheek is likely to be for both parties.

• **Take the occasion and setting into consideration.** Kisses exchanged in the context of business gatherings that are social in nature may be acceptable. However, those exchanged in pure business settings—such as offices, conference rooms, or lunch meetings at restaurants—are usually inappropriate.

• **Avoid kissing up or down the ranks.** If you're the more junior person, bestowing a peck on the cheek of a senior-level executive may be interpreted as currying favor. Doing the opposite can suggest that you're taking advantage of your higher position to make inappropriate advances toward a subordinate.

- **Don't be kissed unless you want to be.** To avoid unwanted kisses, your best bet is to extend your hand before the kisser has a chance to get too close. Another way is to just step back or turn your face slightly.

Women, please note: If you're alone, firmly but politely tell the too-aggressive kisser that his overture was out of place.

- **Don't obsess if your kiss isn't reciprocated.** The person may have been taken off guard and not known what to do. But be sure to make a mental note of his or her behavior so you know whether a future kiss is likely to be regarded favorably.

If you kiss someone and the person reacts negatively—through body language or a comment—simply say, "I didn't mean to offend you" immediately afterward if you're alone; or if others are present, when you have a private moment to apologize.

- **Don't worry about misses.** When a brush on the cheek is the intention, but a lip-to-lip kiss occurs, it's best to just laugh it off.

- **Don't feel obligated to give or receive kisses.** Respect your own comfort, even if you work in a business or city in which kisses are readily exchanged. Give an advance signal by keeping your distance when greetings are being exchanged.

Remember, good feeling can be sent with a smile and a phrase like "It's great to see you" or "I've been looking forward to this lunch."

Male/Female Protocol at the Office

Many male and female executives feel uneasy about office protocol. They're not sure what's appropriate. The male often feels he is between the proverbial rock and a hard place. His first inclination may be to open the door for a woman, as he was traditionally taught to do. But this kind of action may be in conflict with treating his female counterpart as an equal. His attentiveness to a courtesy that he feels is proper may be resented.

The female executive has her own set of predicaments that often lead to uncomfortable feelings. Should she accept special courtesies just for being a woman? In a business setting she prefers to be treated like anyone else. Is she a nuisance to her male counterparts when approaching a door, entering an elevator, or taking off her coat? If she reaches for a door when men are present, is she acting in poor taste? Or when she allows a male executive to open a door is she asking for special treatment?

Believe it or not, there are ways to eliminate these kinds of dilemmas. Basically, the resolution of male/female protocol in business can easily be accomplished with The Offer and Refusal Technique and The Understanding Strategy.

The Offer and Refusal Technique

This approach is simple. It works well because everyone involved is put at ease. The male executive continues to offer the kind of manners he was taught and with which he is comfortable. The female executive accepts those gestures that she believes to be proper in a business setting, but gracefully declines attentiveness she would rather do without. For example:

—When approaching a door, the female executive can slow her pace and allow the man to open the door for her or, if she gets to the door first, open it for everyone else. Should a male reach out for the door from behind her, she can say, "Thank you. I've got it."

—Upon entering a meeting room, if she notices the men beginning to rise from their seats, she can acknowledge the show of courtesy with a smile, or say, "Please keep your seats. Thank you."

—When she is taking off her coat and a man offers to help, she can accept the assistance willingly or simply say, "Thank you, I can handle it myself."

Most men will immediately respond to the woman's wishes, even if it is counter to their own notions of behavior. If the woman declines politely, most male executives will accept the woman's wishes and make a note of her preferences. This leads to The Understanding Strategy.

The Understanding Strategy

Most people who know someone's preferences in terms of business protocol are happy to accommodate. Ultimately, male and female executives who attend the same meetings or travel together will observe each other's preferences and comply with them even though they are not stated outright. In a more formal setting, where many people are present, the male executive might feel obligated to engage in the kind of chivalry he normally avoids. This is understandable. Under these circumstances the female executive can either accept or decline in a gracious manner. In these cases, everyone does what he or she views as appropriate and no one is made to feel uncomfortable.

There are a few situations that are not covered by the techniques and strategies given:

—When a woman and man are having a business luncheon or dinner, the bill is usually placed in front of the person whom the waiter believes to be the host or hostess, based upon his or her actions. More often than not, the bill will be placed in the middle of the table. If it is placed in front of the man and the woman is paying the bill, she should reach over and take it without comment.

—Executives staying at hotels might be better advised to hold their meetings in public rooms rather than the privacy of a hotel room. In some instances a suite offers a sitting room, which is certainly suitable as long as each individual is comfortable with the setting. Good manners and protocol require that the person who calls the meeting be considerate of how others might feel about the meeting's location.

—There are still some private business clubs that restrict women (as our manager in Example Three discovered). To avoid embarrassment, it is wise to check on this before setting a meeting at a private club whose rules are not known to you.

—Male executives should try to avoid always favoring a female executive as minutes secretary of a meeting. On the other hand, unless this practice is abused, female executives should not resent the role just because they are female.

—Female and male executives should take turns preparing and serving coffee, unless all concerned prefer otherwise.

—A male executive addressing a female executive as "Honey," "Sweetheart," or "Dear," is engaging in a demeaning practice.

Business Relationships Outside the Office

N ow that you've mastered business etiquette, it's time to put what you've learned to work. Meeting people outside the office is a valuable way to increase your business network. In this regard, women and men can learn from each other.

By watching male executives, women can learn that some of the traditional methods for making contacts— clubs, sports, college alumni associations—still work. On the other hand, men can take a page from women's book, too. Historically, women have had to socialize to survive. Think about the CEO's wife who has spent a lifetime moving from city to city, constantly making new friends at school, church, and in the neighborhood. That's networking at its very finest. It's no surprise then that women executives in particular have demonstrated a natural talent for networking.

But regardless of gender, all executives must begin at the beginning: Where to find these contacts who will bring you lucrative business, keep you posted on job openings and other grapevine developments, and even offer a boost to the next rung on your career ladder?

Increasingly, women are showing up on the golf course,

a traditional place for meeting important people. (Warning: Don't show up on the golf course without being able to play a passable game. Most clubs have pros who give lessons, and it is advisable to take several before teeing off in front of an important client.)

In addition, health and fitness clubs (preferably co-ed ones), are a good place to work up a sweat and work out a deal. Particularly if you join an exercise class where you see the same people regularly over a long period of time, you can get to know them, what they do, and how their businesses might interface with your own.

Health clubs may be the latest rage, but they can't match trade and professional societies for value. Here you have a head start: Everyone you encounter is in your field. This creates an ideal—and unforced—environment for job-finding and business development.

Women shouldn't forget (and men should take note of) some of the organizations they have used to meet people outside of work. If you have children, getting to know the parents of the other children may turn up someone who will be a valuable source. Similarly, doing volunteer or pro bono work for a community or religious organization may find you shoulder to shoulder with someone you've been wanting to call on.

However you make contacts, remember that the process of networking is meant to be subtle. (That's why formal networking sessions where invitations are forced can be artificial.) You want to get to know people gradually, like making friends or gaining trust in a peer. This takes time, at least six months to a year. For that reason, it makes sense to join an activity you enjoy for your own sake. If you're bored, you'll be unlikely to stick around.

The Role Friendships Play In Your Networking

Don't become so busy networking and meeting new people, that you forget your older friends. Again, here is an area where women have excelled: They are better at nurturing their friendships. Since women have been moving into the workplace, they have rediscovered the value of these ties. Friendships are vital to executives. The stress level that people encounter at work needs a balance. That balance can be achieved through relationships that don't communicate entirely on a work-related level.

In addition, friends can help your business by providing you with future sources, business, and other information that you would not be able to get elsewhere. Many executives, however, find it difficult—if not impossible—to maintain friendships.

Executives feel the crunch of time and may reschedule a lunch with a friend because it is the only expendable item on a crowded calendar. The friend who is put off frequently may decide to call it quits.

Here are some suggestions for maintaining friendships:

• **Make the time.** Realize that relaxing with an old friend isn't a luxury; it's a necessity. That time should include catching up, as well as making future plans.

• **Take turns initiating the meetings.** The phone rings

two ways. Someone who always calls you may begin to feel he or she is forcing you into the relationship.

• **Keep in touch in other ways.** When a time-consuming special project prevents you from leaving the office, send a note or make a phone call to let your friend know you still care.

• **Be creative in finding time for friends.** Arrange meetings at the health club, running track, shopping mall, or wherever your weekend errands will take you.

• **Learn to handle success—both your own and others'.** Rejoice in your friends' success, even if you are having trouble reaching your goals. If the situation is reversed, avoid flaunting your new accomplishments.

• **Stress old friendships.** New friends from work or your health club have a lot to offer, but it is through long-term relationships that you will derive the understanding and nurturing you need.

• **Invite spouses.** If you find it awkward to arrange a get-together with a business friend who is also a member of the opposite sex, make it a foursome and invite your spouses or dates. Who knows? Your friend's spouse (or date) may prove to be someone you should know.

Socializing With Colleagues

You may work with friends and so may find it enjoyable to go out for a drink after work just to catch up outside the office. What happens, however, to those colleagues who are strictly work friends? Should you socialize with them after work?

Yes. The fact is, two or three hours a week spent socializing with your colleagues can bring advancement you'll never earn by working extra hours in your office.

Work takes many forms, and getting together with your boss or colleagues permits open, nonthreatening communication. After hours is when workers trade information about how to handle a client or boss, or about changes on the horizon—vital knowledge that you're not going to hear in company corridors or see written in memos.

How Much Is Enough?

It could be as little as two or three hours a month, or as frequent as weekly outings with colleagues. It has a lot to do with your corporate culture.

Young, fast-paced companies are encouraging after-hours socializing. This may be due to the influence of Japanese corporate culture, in which people regularly retire to a bar or restaurant to talk shop. The value of this vital ritual is now being understood in America.

Play the Game

If you're new to after-hours affability, it may seem uncomfortable at first, particularly if you're a woman and the staff's favorite watering hole is a seedy bar near your office. Give it a try anyway. Some tips:

• **Join the team.** Even if there are pressing demands on your time, go out with colleagues or drop by their gathering places. You don't have to arrive first or stay long. Just make your presence known and make a contribution. It's important for your colleagues to think of you as a team player.

• **Avoid bad habits.** Although the atmosphere is more casual than in the office, don't violate confidences or engage in malicious gossip. Watch physical contact with colleagues of the opposite sex. Keep your drinking under control. You can order a soft drink and still be gregarious— the point is that you participate in the festivities.

• **Be alert.** Your colleagues have a varied pool of experience among them. Socializing is a great way to keep current about developments outside your company and do some valuable networking.

• **Invest the time.** Don't think that attending office parties and other planned activities is enough. The level of communication at these events is never as free as at impromptu gatherings.

• **Take the initiative.** If you're the manager, arrange regular get-togethers for your staff. Particularly if you're a woman manager and find the current gathering place unacceptable, you can select the time and place. Don't force the issue, however. If your staff prefers one place, don't try to change the location.

When You Socialize With Your Boss

Your boss has asked you to lunch, dinner or a weekend golf game. If it's the first time or a rare event, you may feel anxious because you're unsure what the purpose is or how to act. But if handled correctly, these occasions can help you develop a stronger working relationship with your boss. Here's what to keep in mind:

• **Take cues from your boss.** If your boss keeps business out of the conversation, it's risky for you to mention office matters. If you're unsure whether something is appropriate, refer to it briefly and watch the reaction. Nonverbal behavior (sitting back, looking away) can be a clue that your boss doesn't feel comfortable talking about the subject. So can a verbal response—a humorous comment rather than a question.

• **Drink in moderation—or not at all.** Even if a few drinks don't normally affect you, it's smart to stick to a wine spritzer or a beer if you feel it's appropriate because your boss is drinking. Drinking more isn't worth the risk of saying or doing something you'll later regret.

• **Avoid revealing personal information.** No matter what your boss chooses to tell you about his or her personal affairs, it's not smart to talk about your marital, family, financial or health problems. Nor is it advisable to betray any lack of confidence about your professional abilities, your career choice or your job. What your boss doesn't know really can't hurt you.

• **Be cautious in your comments about other staff members.** If you're asked your opinion of other employees or what you know about them personally, don't be overly critical or supply too many details. Instead, try to address your boss's concern and offer suggestions on how the problem or situation could be remedied. If, for example, you're asked whether you've noticed that a colleague has a drinking problem, talk about ways the person might be helped rather than focusing on the times he or she has been late or botched an assignment.

• **Keep the socializing "professional" if your boss is of the opposite sex.** The more relaxed the setting, the more opportunity there is for misinterpretation of language and behavior. It is wise to decline an invitation if you feel the place or function is inappropriate for a working relationship. Individuals who suspect their boss's intention may be romantic or sexual are smart to bring up their spouses, "significant others" or family if the conversation takes such a turn.

• **Don't confuse your role as a subordinate with your role as a friend.** Even if you and your boss become buddies, you shouldn't lose sight of the fact that your boss still has power over your job and your paycheck. And if you expect special consideration because you're a friend, you're setting yourself up for disappointment.

Recommendation: If your boss invites you more often than others in your office, don't advertise that fact. Your colleagues may be reluctant to share information or work closely with you if they feel you have access to the boss that they don't.

Section Two
YOUR WORKING RELATIONSHIPS

Male-female relationships in the workplace have come a long way since the 1950s when men went to work and women stayed at home.

• **Male and female managers** are generally feeling more positive about working and competing with peers of the opposite sex, learning to accept it as part of today's business landscape. Problems often arise, however, when a man ends up working for a female boss—still an unusual situation in most companies.

• **Women in middle management** seem to be getting over the feeling that they shouldn't compete with men. Younger women compete quite effectively, possibly because they grew up with the idea that they would pursue careers.

• **Women who compete with women** often experience difficulties. Because some obstacles for women to move into senior management positions still exist, problems between female colleagues may be more visible than those between men and women.

Before we begin this discussion, be honest in sizing up your own attitudes about working with and competing with a member of the opposite sex. If you're experiencing difficulties, talk about your discomfort with a colleague you can trust—someone of the same or opposite sex, whomever you feel more comfortable with. Also, talk to people who have been in similar situations.

If you are behind the times and believe that women belong in the home, remember in this labor-short climate, only the fullest development of all human resources will result in any company's success. This depends on every person's ability to cooperate—and also to compete—with a high degree of effectiveness.

Also, remember that managers who display "people skills"—that is, the ability to work with and supervise a diversified work force made up of women, minorities, the handicapped, and older workers—will be extremely valuable to their companies in the coming decade. Overcoming any biases *now* will ensure your advancement *later*.

The Masculine Mystique About Managing Women

Even the most fair-minded male executives are some-times guilty of unintentional discrimination against women on their staff, according to a study undertaken by Catalyst, a New York-based national research and advisory organization. The reason: Male managers' perceptions of women are often based on traditional cultural values rather than current work and family realities.

One result is that men and women feel uncomfortable dealing with each other, and that can lead to communication and productivity problems. Here are some of the most typical situations in which men's well-intentioned behavior can backfire and how you can reduce the possibility of having it affect you and your female staff members:

• **Awarding promotions and relocating.** Given two equally qualified people, one male and one female, managers often assume that the male is the better risk because he is more likely to put his career ambitions ahead of any family considerations. That assumption is not necessarily valid. There are many women who will make sacrifices for their careers and men who will put their families ahead of their work.

Before you rule out a female candidate, make sure that you know what her ambitions and level of commitment are.

If women who work for you haven't articulated their priorities, ask about them. It's also important that your female staff members be aware of what's necessary for advancement so that they can decide what they are—and are not—willing to do.

• **Performance feedback.** Some managers find it difficult to give women honest criticism because they're worried that women will react emotionally—and perhaps even cry. Realize that unless a woman knows what her weaknesses are, she won't be able to improve; so you're effectively denying her the right to advance. (More on dealing with an employee who cries appears later in this Book.)

• **Travel.** When overnight or extensive travel is required, managers often send male staff members because they assume—often subconsciously—that mothers should be at home with their children and wives with their husbands, or that travel isn't safe for single women. These assumptions work against female staff members because travel can provide valuable job experience. Again, the best policy is to ask women on your staff what their preferences are and to let them know the type of experience they might gain before you rule out the possibility of sending them on business trips.

To find out if your polite and proper behavior may in fact be resented by the women you manage, you need to analyze your own reactions to different situations. Talk to female colleagues and staff members to learn about their perceptions of your attitudes.

If you find that you do treat women you manage differently, that insight alone may change things. It will certainly make you more sensitive to your own prejudices and how they affect others. And you probably will consider your decisions about female staff members more carefully.

Making the Occasional Exception

The male manager should strive to keep his management style nonsexist. Yet, there are occasions when he will have to make exceptions to the rule. For example, the experienced male manager might want to give some extra assistance to a recently promoted female manager.

Why make an exception? A woman manager often encounters malice, jealously, and sabotage from colleagues, subordinates, or superiors who resent a woman who "doesn't know her place." And if she complains about it, the hostility is likely to increase, as well as the despised feelings toward her.

The male manager shouldn't hasten to clear the air, though. Chances are that she won't want him to intervene in the situation, but as her mentor there may be times when he must. It is necessary to prevent bias from undermining their authority. His intervention might be necessary, for instance, to:

• **Guide her through political minefields.** The aggressive, up-and-coming woman may need *aggressive* intervention. Her head-on charges into male-dominated bastions could upset the very executives whose support she'll need one day to realize her ambitions.

• **Reinforce her authority.** If the female manager demonstrates a lack of assertiveness that is undermining her effectiveness, an experienced executive must step in and take control. *Example:* If she does her own typing because your clerical staff claims it doesn't have time to do her work, tell them that *you* object to the situation: Managers make expensive typists.

• **Get her to take due credit.** Some new women managers haven't yet learned that they have to toot their own horns. They may refuse to take credit for excellent work or blame themselves for a team failure. She needs to be told that she can't continue to underrate herself.

• **Don't let others put her down.** Most likely, she neither wants nor needs special protection. But if nothing else, her boss can strongly object when others attribute her successes to "luck," "feminine intuition," or "good looks."

Developing Your Management Style

As you can see, developing a management style that is effective requires a combination of time, testing, and patience. You want to establish the right atmosphere and tone. You don't want to come across as a tyrant, but also you don't want to seem to be too easy going, too informal.

Is there a fine line you should walk between the two? You want to get along well with people, but you also have to make sure that the work gets done efficiently and effectively.

Organizational style, these days, tends to be more informal—which is usually an advantage: The atmosphere is more relaxed, communication barriers are reduced, problems surface more quickly and easily. Concern and cooperation are also more readily offered.

Informality can be stretched too far, however, and this can cause difficulties for the person in a supervisory position. People may ask for special favors, for example, or shrug off criticism. The atmosphere may become so relaxed that productivity starts to slow down. (Women managers are at risk in that informality may be interpreted as weakness).

The question, then, for anyone in a managerial position is how can you gain the advantages of an informal style while not crossing over into too much informality? (Essentially, what we are talking about is a combination of the best masculine and feminine traits of management style). Here are some answers:

• **Proceed with caution.** For anyone who is just taking over supervisory responsibilities, the best approach is to move slowly and cautiously. Give the situation a little time, assess the people you will be working with, be alert to the signals they send when you question them, assign work, and so on. And lean toward a more formal approach as you're sizing things up. It's a lot easier to go from formal to informal than it is the other way around.

• **Remember that you don't have to be a friend.** This is something that people (particularly women) who are new to managerial responsibilities often fail to take into consideration. In their desire to be liked, they go too far with the friendship approach, and this ultimately hampers their ability to give directions, or offer justifiable criticism, or make unpleasant decisions.

Actually, trying to act as if everyone is your friend won't help you do your job any better. A relationship can be businesslike and matter of fact—and still be productive.

• **Let work be the primary topic of conversation.** This doesn't rule out casual conversation, of course—that's an essential part of building a relaxed atmosphere. But your main purpose, after all, is to make sure that the work gets out. So don't hesitate to tell employees what's expected of them, explain procedures, ask questions, share organizational news, listen to problems, be ready with encouragement when the going gets tough. All this can be done informally, but in a way that clearly indicates that work is the number one concern.

• **Be comfortable in your role.** Undoubtedly, there will be times when you have to use pressure or exert your authority. You will have to criticize, evaluate performance, make a decision that is difficult for others to accept. But if you have been fair and even-handed in the past, your remarks and decisions are likely to be accepted, in spite of any initial grumbling that takes place.

Display a nonsexist attitude. Treat the women on your staff the same way you treat the men. You should not discriminate for or against anyone based on his or her sex. If you show concern for a woman's child care arrangements during a business trip, do so for the men on your staff, too. Demonstrate that you will judge employees on their performance, not on their sex.

■ **Observation:** One of the benefits of a more informal style is that in an open, relaxed atmosphere, you are much more in touch with what is really going on. People feel free to discuss problems. They are also quicker to propose ideas of their own and more receptive to suggestions from you. Basically, it's a matter of treating people like adults—so that they respond in kind.

Be Tough But Be Fair

A good manager is a tough manager. Unfortunately, too many managers interpret tough in terms that could best be described as "macho male"—shouting, swearing, intimidating, threatening. The influx of women into the workplace has not changed this fact. Many male managers who feel threatened by ambitious women subordinates attempt to throw them off guard by using these guerrilla tactics. At the same time, many women believe they must exhibit these aggressive male characteristics to succeed.

Yet there are many executives—men and women—who are labelled tough but who seldom inspire fear and anxiety. They are demanding in the goals they set—and you probably wouldn't work for them long if you didn't meet those goals. They get their reputations for toughness, however, from one or more of the following:

• **They insist people have the facts.** They tolerate dissent, but only when it's backed up by careful research and arguments that can't be shot down. "Go do your homework," they will say if subordinates don't come up with well-informed recommendations. They're tough all right, but hard to fault.

• **They treat everyone equally.** Some find it hard to keep themselves from growling at mistakes, but at least they growl at everyone. Many bad-tough bosses are really no more than bullies—they'll snap at people who cower easily, but treat with respect the ones who don't. Good-tough bosses either yell at everyone or, even better, roar only at those who usually roar back.

• **They go to bat for their subordinates.** They expect more of people than most of their colleagues do, then they fight hard to get them the resources they need to do a first-rate job and the rewards they deserve for top performance.

• **They demonstrate their personal concern.** Tough

may often equal rough but even this trait can be mitigated if the boss also gives signs of caring about people. It's possible to be demanding and still drop into the office of a subordinate just out of the hospital, for example, to fill the person in on recent developments and say, "I want you to take it easy until you feel better." It's also possible to criticize a subordinate and still come back an hour later and ask, "Has your daughter come back from Mexico yet?" Good-tough bosses don't nurse grudges and they let everyone know that, however strict their standards, they care about people as much as the work.

- **Observation:** Taking the "macho" route to earn a tough reputation means that everyone snaps to attention when the whip is cracked. But it also means creating a team that's secretly waiting for the boss's failure, rather than doing all it can to ensure success.

The Iron Fist In a Velvet Glove

There are alternatives to the tough management style. In fact, many female managers are coming to realize that the supposedly "feminine" traits of compassion and accommodation, and the talent for managing in crisis can be combined with firm direction and efficiency.

Compromise and compliance is the management doctrine that numerous companies are adopting today, modeled after so-called feminine traits. It seems to be one of the oldest psychological techniques in the world and it works as well in negotiations as it does in any other form of human relations.

Another feminine approach is the concept of "employee empowerment," allowing employees to make more decisions about their work. This tactic helps to stir people to produce their best because they feel good about themselves and the place where they work.

Here's some advice polled from successful female executives:

• **Don't be afraid to share glory with subordinates or supervisors.** When you give away power, you enhance your own standing in the eyes of colleagues.

• **Attributes of a good mother often apply to good management.** The ability to comfort, praise, scold, entertain, teach, punish, reward, and support help to shape others' behavior.

• **In high-pressure situations, fight the tendency to respond emotionally.** Use humor instead; it usually defuses tension and hostility.

Establishing Relationships With Those Above

A mentoring relationship is actually a close political alliance in which the flow of information goes more strongly in one direction than in the other. This usually occurs when a more senior person recognizes some special potential in a subordinate and singles him or her out for cultivation.

Unfortunately, women have not had the same opportunities to develop mentor relationships. There are still not enough high-level women who can serve as mentors for younger women. Many of the famous older-men/younger-women mentoring situations we have heard about at major corporations resulted in romantic liaisons, raising the idea that it is impossible for a man and woman to have a purely business-like mentoring relationship.

In fact, the entire concept of mentoring may be overrated. A broad network of cultivated relationships is likely to be more beneficial to a younger manager than a strong political alliance with one senior person. However, if there is one powerful person who seems to take a special interest in you and wants to further your ideas, the advantages of cultivating a relationship may outweigh the lack of political mobility you'll have to accept.

What a Well-Positioned Mentor Can—And Cannot—Do for You

A well-placed mentor will speed your progress and, with luck, teach you some important things. However, you should be aware of the following pitfalls:

• **Picking the wrong mentor.** In your eagerness to establish ties with a bigwig, don't establish ties with somebody who is not well connected, who doesn't know as much as you expected, or who is about to leave. Do your homework carefully before establishing ties. You're better off being unallied than tied to the wrong person or camp.

• **Getting hamstrung.** When you have a mentor, you can worry too much about taking risks out of fear that your actions will reflect badly on your ally. So you become an ineffectual clone who waits for suggestions of what to do. *Defense:* Try to find a mentor who encourages independent efforts, and continue to act with autonomy in most of your projects.

• **Getting reined in.** You have an idea you think is terrific, and go ahead and propose it at a meeting. To your surprise, your mentor is furious because he or she wasn't consulted or thinks it is a bad idea. *Problem:* When you agree to a mentoring relationship, you are also agreeing to a certain amount of control.

• **Political limitations.** Having a mentor not only cuts you off from having close ties to your mentor's opponents, it also creates disturbances with colleagues at your own level. How would you like to have a colleague who shares your rank and title, but who has the ear of top management while you do not? Of course, your colleagues are free to advance themselves politically, but you must calculate the risk of schisms with your co-workers.

• **Ugly surprises.** Just because your mentor tells you all about certain things doesn't mean he or she will tell you you're getting bypassed for a promotion, or that a key assignment is going to somebody else. The biggest surprise of all is when your mentor leaves. Never forget that you are a subordinate, and there are limits on the information you'll receive.

• **Blurring of roles.** Always remember that your mentor is your superior. In almost all cases, it is unwise to confide that you are unhappy with your job, or that you are seeking a new job. Try also to avoid gossip, unless a mentor asks you for it specifically. *Example:* Your mentor asks, "What is the feeling among your colleagues about the new management development program?" In matters of joke-telling, sharing personal information, and social interaction, always let your mentor take the lead in establishing guidelines. For example, never be the first to suggest that you have lunch.

• **Burn-out.** A close relationship can eventually degrade if you become a threat—or a mental burden—to the person who has taken you on. Be willing to distance yourself when necessary and don't overtax the relationship by asking for advice too often or stopping by when you have nothing—or something minor—to discuss.

• **Dangerous liaisons.** Make sure that your mentor relationship is one based on business and not on sexual attraction. If you find yourself sexually attracted to a possible mentor, it might not be a good situation for you. Also, be wary if your potential mentor has left in his or her wake a string of attractive, young executives whose careers were not furthered (and possibly were damaged) by the relationship.

How to Make Points With Your Boss

One way to keep your relationship with a mentor or with your boss in line is to follow certain guidelines yourself. Your relationship with your boss is likely to be taken as a microcosm of your future with the company. If you manage it well, it is taken as a barometer of your potential in the firm. Here are some vital pointers to managing this all-important relationship effectively:

• **Question your viewpoint.** If you think your boss is either beyond reproach or incapable of doing anything right, you have failed to make an accurate—or useful—assessment of his or her abilities. And you are not meeting his or her needs. Go back to square one and take an objective inventory of what he or she can do.

• **Learn how to format information.** As a subordinate, one of your major duties will be to supply your boss with facts. Don't format the process the way *you* like it. Try to meet your boss's preferences. If he or she prefers a daily morning chat to a stack of memos, oblige that need.

• **Understand his or her personality—and your own.** For the relationship to work effectively, you may have to subjugate your own style. To accomplish this, you will have to do some self-analysis. If you're a fun-loving individual and your boss is glum, superimposing your style on the relationship can result in disaster.

• **Learn when to make proposals—and when to keep quiet.** Especially in the early stages of a relationship, it is vital that you offer only your best, and most fully developed ideas. Spattering your boss with a buckshot hail of suggestions in the hopes that some will hit the mark only brands you as someone who is right two or three percent of the time.

• **Anticipate pressures.** Pay attention to how and when your boss feels pressure most acutely. *Examples:* Is it when he or she is facing a meeting with a particular high-ranking person? Is it when he or she prepares reports? Try to throw your support at these areas first. Don't blindly believe that your priorities match those of your boss.

• **Defend your boss.** Never gossip about your boss's shortcomings—it's far too dangerous a game. If your boss has fallen on hard times with upper management, then you will have to make an appraisal of his or her situation and act accordingly. In all but the most extreme cases, defending your boss is your own best defense.

Management Style—Dealing With Special Situations

O nce you have established your management style, you will set a certain tone for your working relationships. Your colleagues, observing how you treat others, will understand how you expect to be treated by them. Yet as you probably already know, you must expect the unexpected. Just when you think you know how to handle anything thrown your way, a new situation surfaces and throws you off balance. Pondering these unusual events beforehand will help you devise a method for coping.

Fear of Crying—Handling an Employee's Tears

How do you handle someone as an adult when he or she is acting like a child? A male manager is criticizing a female employee's work, or telling her she's not going to get the raise or the promotion she wanted, and suddenly she's in tears. What does he do?

The crying episode is an awkward, stressful situation that may make the manager reluctant to give straight feedback to his employee in the future. Fortunately, though, the strides women have made in leadership roles have decreased the frequency of episodes in which they cry; obviously they don't want to be seen as a manager who's likely to break down and cry in public. Ironically, men are showing tears more often. In spite of these role reversals, society still applies a separate set of standards to men and women who show their emotions openly.

The Male-Female Double Standard

Even though studies show that women cry about four times as often as men, there is no evidence that they cry more frequently *at work*. Yet many managers—especially males—cannot handle the situation when it does occur because of their own emotional responses.

For men, outbursts of anger have traditionally been a more socially acceptable release of emotion than tears. A male manager may therefore be caught off-guard and not know how to deal with a crying man.

A woman manager may find the situation tricky because the crying employee—man or woman—will expect comfort from her, viewing her as a "mother" figure.

Why is it so difficult for managers to cope with an employee's tears? The two main reasons are:

• **Prejudiced attitudes.** The boss sees crying as weakness or a plea for help even though it may be a release of simple frustration or anger. But a manager may misunderstand and feel manipulated.

• **Role confusion.** Rather than treating the situation as

a manager, the supervisor may treat the crying episode as a parent; he or she feels obligated to listen and help the worker calm down. Afterward, the manager may feel that he or she is expected to treat the employee as a supervisor *and* as a parent.

Whenever you have bad news to report, give some thought as to where and when you will meet. Select a private place and a time when you won't be interrupted. In addition, here are some guidelines for dealing with a tearful employee:

• **Maintain your professional relationship.** Don't act like a parent or a friend and become overly solicitous. Let the employee use his or her own resources to regain the appropriate composure required for a working atmosphere.

• **Stand your ground.** Make it clear that you are criticizing the problem at hand and not the person in particular. Don't associate the problem with the employee. Eventually work out a structured plan that will prevent a recurrence of the problem or improve the situation.

• **Bring in a buffer.** Often, crying comes as a complete surprise to the manager. But if you know that the person is highly sensitive, arrange to have a respected staff member sit in on the conversation.

• **Know when to stop.** If a worker cannot retain his or her composure or the session becomes hostile, it's time to stop. The best action is to end the conversation and resume it when emotions have cooled down and the employee can function normally.

• **Don't make rash judgments.** A single incident of crying should not be taken to mean that an employee is too emotional and cannot handle the job.

■ **Observation:** Open communication remains the top priority. If managers avoid criticizing their employees for fear that they may cry, both parties will suffer from the lack of communication. The employee may see lack of feedback as an indication that his or her talents and abilities are insufficient. This will result in low morale and impaired productivity, and it may leave an employee no other choice but to resign.

Helping Employees Through Personal Crises

Often an employee may cry not because of something you said but because he or she is in the throes of a serious personal crisis. There's a natural urge to sympathize. That urge makes good business sense because it strengthens a humane working relationship and helps the employee to recover.

Sympathy is an emotion that is easy to feel, yet often difficult to convey. Consider this advice:

• **Listen carefully.** Let the employee express feelings (anger, sorrow, frustration or disappointment) without being judgmental. Avoid hindsight reflections that focus on "what might have been."

• **Help the person to accept reality.** It may seem comforting to minimize or deny the impact of the trouble. To help the employee accept what has happened, you are in a position to provide a helpful perspective, especially if you suffered a similar personal crisis, or if the crisis is work related. However, don't offer hopes that may prove false. There can be nothing more annoying to someone in pain than to be fed optimistic clichés such as: "A month from now you'll see it all in a much clearer light," or "You've got to take hold of yourself."

• **Remind the employee of his or her strengths.** At times of loss, people's self-confidence often flags under stress. As manager, you are in an excellent position to address the employee's personal qualities that can be called on to weather the crisis.

• **Offer specific aid, if possible.** In cases of personal crises, remind the employee of the company's policy for paid time off or extended unpaid leave. If the situation calls for it, review relevant health-care benefits. Check with your personnel officer for any benefits you may have missed. For work-related crises, offer the person assignments, or further training, that can help him or her recover professionally.

■ **Observation:** Asking the employee to reveal all the personal details of the trouble, or his or her plans or intentions, can put you in an unwanted guardian's role. Don't push beyond what the employee is willing to volunteer unless it involves business matters for which you intend to take full responsibility. Anything else is both a personal and private matter.

The Special Needs of the Two Career Couple

—A single father refuses overtime in order to get home to his children.

—A pregnant employee wants to work only part time after her baby is born.

—A vice president has a child-care crisis and cannot attend an important meeting until she can arrange for a substitute babysitter.

Fewer and fewer families today follow the traditional pattern of the wife staying at home and the husband working to support them both. This change has come about partly because of women's increasing acceptance in the work force, and partly because of economic necessity. Many families say they can't afford to live on just one salary.

As a manager, it is your job to be aware of the needs of each member of your staff. Don't resort to old stereotypes. It is incorrect to assume that a married man in your employ will more readily agree to relocate or work overtime than a married woman. Many men today "share" child-care duties with their wives.

You may be faced with the reality of dealing with an employee whose family demands may be affecting his or her performance. Because this area is still a new one, there are no established answers, forcing you and your organization to be creative in coming up with solutions. Here are some practical approaches:

- **Be understanding.** Managers who coolly adopt an attitude that implies, "This is your problem, not mine" are creating more stress, making it that much more difficult for the employee to cope and concentrate on his or her work. Instead, hear the person out. Sometimes all the employee needs is some understanding and emotional support during a trying time.
- **See what company resources are available.** Many companies are adopting policies and fringe benefits to meet contemporary family needs. Find out what your firm offers: an Employee Assistance Program that provides professional counseling for family problems, for instance, or a linkup to local child day-care programs, or alternate work schedules, such as flextime.
- **See what leeway you have.** Even if your company has no policy that meets the needs of an individual employee, you may still offer help. You might be able to develop a schedule that allows a parent to come in and leave an hour earlier or later than everyone else, or work at home during a child's illness.
- **Consider the duration of the problem.** Analyze whether a particular problem is a temporary emergency (like a sick babysitter) or a lengthy one (like an employee with a terminally ill parent). You can make ad hoc arrangements for short-term problems, but for longer-term ones, you may need advice or permission from the higher-ups.
- **Consider the impact on others.** Fearing that a precedent will be set, many managers hesitate to make special arrangements for one employee who is struggling. Consider the other side, however: When a good employee is

treated like an individual, others will know you care.

- **Observation:** If you too are subject to work-family conflicts, you know the difference it makes to have an understanding boss. Put yourself in your subordinates' shoes and consider all sides of a problem before taking action. You might even want to go a step further, and use your influence to encourage the establishment of supportive policies throughout your company.

Men vs. Women: The Public Insult

When the business office was a male-only environment, disagreements could be settled "between men." Circumstances have changed. A comment made man-to-man might have been permissible. When delivered man-to-woman or woman-to-man, that same comment may have implications—and repercussions—that could damage the reputations and careers of the participants.

What rules govern your specific workplace will depend on many factors, including how many men and women are on your team. Keep in mind, however, that adding one new person to your staff may be enough to tip the balance in another direction. That person may have different expectations of how people should behave in the work setting. Now, more than ever before, it is prudent to think before you speak and to consider the full impact of your remarks.

Refrain from using "bad" language. An occasional salty word may not get you transferred to Siberia, but it won't advance your career either (unless your boss is one who swears like a sailor and expects his staff to do likewise). In general, most four-letter words are so overused these days that they have lost their ability to shock. And, since you never know who will take offense (men or women), it's best to omit these words from your vocabulary.

What happens if you are insulted or embarrassed by a manager of the opposite sex in a public meeting? You shouldn't let the gender of the offender influence your reaction. The person needs to be dealt with swiftly and sternly.

Consider the following example:

Patricia Quigley listens to a lengthy point made by another manager in the staff meeting. Confused, she says, "At the risk of sounding stupid..." The other manager (a man) smiles sardonically and fires back, "No risk at all, Pat. If you don't understand, then you must be stupid."

The retort is strained, the laughter is sparse—and Pat is left speechless by the put-down.

Here are some possible courses of action for Pat, and for you if you ever find yourself in such a situation:

Have It Out Publicly

Although this type of action should not be your first choice, it can be remarkably effective if done properly. The problem is that your open, aggressive response could cause much discomfort—especially if the group has tolerated put-downs to others. Thus, your challenge may result in an attempt by one or more colleagues to move on to another, less controversial subject. Then, in effect, you will have been shut up twice.

Although aware of the possible consequences, you may still be willing to risk yourself in public leveling with your colleague. To do so with some degree of success:

1. Don't join in the laughter. When you do, you are sanctioning, or seeming to sanction, what has just happened to you.

2. Don't hide the fact that you disapprove of the put-down, but don't lose your temper, either. Undoubtedly, other people are already embarrassed, and your rage may push them to the limits of tolerance. They may try to isolate you, to keep you from being effective.

3. Use the right kind of response to nullify the put-down. For example: "I don't think that was very helpful. I have the feeling I've been put down and I'm wondering why," or "That's a brush-off isn't it, not an answer?"

You have been made the victim—you know it, the put-downer knows it, and everyone else present knows it. Therefore, acknowledge the fact—and act the part. You may cause the put-downer to back off, even to apologize. And it's possible that you may enlist an ally or two among the witnesses.

Have It Out Privately

If you don't want to respond to such rudeness in public, you might arrange to have a private talk with the "offender." Again, try not to lose your temper. Describe exactly how you felt when you were put down. Don't accuse the other person of trying to embarrass you or shut you off, or of acting with malicious intentions—because he or she can, and probably will, deny that was intended. If you confine yourself to describing how you felt at that time, your adversary can't deny your right to speak up. He or she may try to rationalize, excuse, explain—but your feelings can't be dismissed.

It's possible that you won't get any immediate satisfaction from the meeting. You may not get an apology, either. In fact, the put-downer may even try to embarrass you further to avoid any show of personal discomfort.

When that happens, all you can do is give notice that you are not going to take future put-downs without some kind of response. This alone may discourage any more attempts to embarrass or humiliate you.

Ignore the Put-Down

When the put-down is a one-time occurrence, when it seems to indicate no malice on the part of the other person, when it is probably used just to close out discussion, then you may want to overlook it. This isn't easy to do, of course. We all have our pride. But you may be able to score some points by getting the response you want. Some follow-up comments you might make:

1. "This is still not clear to me. Maybe it is to everyone else and I'm the only one having trouble. If so, I hate to take up the group's time." Put that way, you'll frequently get a constructive response from someone. You may find that others in the group don't have any clearer idea of the point under discussion than you do. Or you may find that you stand alone, but that someone will try to explain—without rancor—what is unclear to you.

2. "I guess I sound stupid, but I don't see the point (or understand the distinction). Can somebody help me?" Put sincerely, this question seldom fails to get what it asks for.

If you can keep your voice level calm, your quiet dignity will go a long way toward disarming the person who tried to put you down. While in most cases everyone present can see a put-down for what it is, that doesn't mean that it has been admired or accepted by others. Your response may have the effect of not only dampening the impact of this put-down, but also of discouraging such behavior in the future.

- **Observation:** The dynamics of the situation change with regard to the gender of the participants. If a man puts down a female employee, he may appear to be a bully. In that situation, the woman will want to react with force to make it clear that she will not be intimidated either then or in the future.

  If a woman is doing the putting down with a man on the receiving end, she may come off as being defensive, resorting to a put-down rather than being on sure footing. In this case, the man would do best to ignore the put-down and press her for the facts.

Show Your Feelings

If you have found yourself in a situation similar to the one above, or been an observer, you know how difficult it can be to choose the best response or course of action. You don't want to put people off by coming across as militant, angry, or a poor sport. On the other hand, you don't want to tolerate those types of behavior that make you feel uneasy.

Unlike social situations, you have another factor to consider in a work environment: office politics. Everything you do or say is bound to reflect on your organization and your superiors. There is also the very real possibility that your behavior can affect your job stability and income.

With all of this in mind, how is it possible to stand up for your dignity, while at the same time protect these rather delicate professional relationships?

Here are some suggestions that may prove helpful:

- **Trust your feelings to show through.** Hostility breeds hostility. So many people make the mistake of acting overly nice. Don't worry about sounding hostile if you feel violated in some way.

- **Frame your concern in a positive way.** If you do feel really angry, there is usually an effective way to communicate the fact that you dislike such treatment. Speak pleasantly but firmly. When someone calls you "Honey," you might say, "I'd prefer to be called Marie, if you don't mind." The other person may become flustered and make an apology, or try to say something humorous. If you feel he or she is not challenging your objection, maintain a pleasant demeanor. But remember that you do not need to apologize for speaking up, nor do you need to explain yourself.

- **Soften it with humor.** Light humor can go a long way in spreading good feelings and replacing negative ones. A male secretary, for example, who objects to worn-out sexist jokes might say something like, "You women had better stop this now, or I'm going to get the Society of Male Secretaries after you." This communicates his dislike without offending anyone or putting anybody on the defensive.

• **Evaluate the total situation.** Do all instances of male-female touching have sexual overtones? Sometimes, when a boss puts his arm around the shoulders of a male subordinate, it is a way of saying to outsiders, "We have solidarity; we are a team." It is possible that a boss who does this with a male subordinate could be doing the same thing with a female employee for the same reason.

It's important, therefore, to consider different possibilities before jumping to conclusions. (Susan, in Example Four, would be wise to restrain from criticizing her superior for hugging her unless the incident is repeated.) If you feel that a particular person may be touching you without any sexual overtones intended, but you still prefer he or she would refrain, the best response is to say something like, "I know you are in the habit of touching people and that you don't intend any harm. It's just that I would feel better if we could conduct our business with only verbal communication."

• **Draw the line.** You may feel so strongly about certain situations that you really don't care how a boss, customer, or client might receive your response. Take the sexist joke told in a meeting. Should you compromise your values by laughing at something you don't find funny? Again, office politics are at play. If the situation is tense it is not wise to add to the tension. Your best move might be to feign distraction. If you feel you can win points by showing your disdain, then do so.

- **Observation:** Don't overlook the role of body language in getting your point across. At the same time, studies show that men may perceive sexual overtones where women don't. Therefore, if you are a woman, use caution. Touching someone can communicate many things: status, friendship, dominance. If you prefer not to touch, you can still communicate power by standing while the other person sits, spreading your papers on the person's desk, or placing an object belonging to you on top of something of his or hers.

Family Feuds: When Staff Members Fight

Not all men like working with women. And not all women work well with men. Does the following case study sound familiar to you?

For as long as anyone can remember, Samantha Oates and Terrence Phillips have worked together with barely suppressed dislike. Some have chalked it up to a personality conflict. You suspect that the problem runs deeper. Samantha has often shown difficulty cooperating with the men in your group. Similarly, Terrence seems more comfortable working with men.

As time goes by, people learn to live with and work around the friction. Yet you, as their manager, don't have to accept the feud as permanent. Obviously, you can't mandate a more cordial relationship, but consider these factors:

• **You're not expecting them to become good friends.** That would be unrealistic. Besides they don't need to be buddies to work together well.

• **You can't expect a fast turnaround.** It's taken a long time to get them where they are now. It will take time to change attitudes. So, expect that the best results for you will come by being deliberate.

• **You'd like to help them change their perceptions.** But you will never be able to sell them on the virtues of working with a member of the opposite sex. You may be more successful convincing them of each other's attributes. And bit by bit you may be able to chip away at the overall prejudices they feel.

• **Talk to each of them about the other.** Just get them comfortable with the idea of hearing you say something like, "Samantha told me an interesting thing the other day...." The effect may be a stiffening at first. But as you continue to refer offhandedly to one subordinate in the other's presence, both will relax after a while.

• **Try eliciting some compliments.** For example, in talking with Samantha: "Did you see that project analysis that Terrence prepared? I thought it was a very professional piece of work." If Samantha agrees, you can then declare, truthfully to Terrence, "Samantha told me she was impressed with your project analysis." A simple statement like that could turn out to be the first icebreaker. Be careful to keep a visible hand in trying to get each to see the best in the other. If you appear to be leaning in one direction, you may intensify the hostility.

• **Put them to work on the same committee.** Include men and women who enjoy working with members of the opposite sex. In this way, they will begin to see what happens when workers—men and women—work together to achieve a common goal.

■ **Observation:** Many times, in such cases, the prejudices may be so deep rooted that opinions will never really change. You may get Samantha and Terrence to work together successfully, but when Samantha is paired with John, and Terrence with Diane, the entire process could be repeated. If you sense a definite pattern, you have no other alternative than to confront Samantha and Terrence with the future implications of their inability to work with others.

Hopefully, that drastic measure won't be necessary. You may find that if you can deal with the issue on a day-to-day basis, stressing that the work at hand, not individual feelings, must take precedence, you will find that, over time, real progress will be made.

Women vs. Women: Women Colleagues Together

Women executives working in groups usually exhibit a high level of trust, productivity, and creativity. However, problems can arise—when women work together in strongly male-dominated firms.

Here are some problems you've probably seen:

• **The "Good Woman-Bad Woman" syndrome.**
Management—or the climate it has fostered—plays one woman against another in the struggle for power. *Example:* The first female manager at a firm has worked long and hard, and has turned around many biases in the process. Impressed with her performance, management hires a second woman, who is subjected to much less demanding treatment and given more respect up front. Resentments may result between the two women.

• **Male managers' sexual attractions.** Even when women laugh at male courting patterns and strive to rise above the friction they cause, problems can frequently result. *Example:* A well-situated male executive flirts with a new woman on his staff, and is overly attentive to her ideas. Women who have been there longer may reasonably expect a lengthy sorting-out period in which they will have to work harder to get their ideas heard. Again, tension can result—despite the fact that the problem was not caused by the new female staff member.

• **Different weighting of family responsibilities.** One woman feels comfortable leaving her children in the care of her husband so that she can work late several days a week. Another woman leaves promptly at 5 p.m. every day. *Problem:* Men have traditionally not been questioned about how they balance their work and family time. In a male-dominated firm, women's individual choices regarding their family schedules become highly visible. With no clear guidelines, women may question their own priorities or those of their colleagues. Again, friction can result.

- **Avoid referring to the more experienced women as "trailblazers."** That designation will offend them and turn off the younger women who feel they are advancing based on their own talents, not because someone cleared a path for them.

At the same time, make it clear to these long-term employees that you value their contributions by soliciting their opinions on important matters.

- **Encourage cooperation among the two camps.** Make it clear that you respect *all* women on your team and expect them to put aside any prejudices they may feel. If you suspect that someone is attempting to disrupt the working environment, have it out with them privately.

- **Avoid sexual liaisons.** No matter how attracted you may be to a female member of your staff, realize that acting on those feelings is bound to create tension and resentment among the other women.

Be aware of any sexual liaisons that may develop between lower-level managers and the people they supervise. These associations could prove to be disruptive and you may need to step in quickly before they get out of hand.

- **Show understanding.** You should expect all members of your staff to arrange adequate child-care arrangements. Offer assistance if you suspect there is a problem. Many companies have referral services that can help employees locate reliable child care. Or you may want to keep a list of local child-care providers that you can pass out to interested employees.

When an employee does have a child-care emergency,

try to be understanding. Provided the crisis isn't long-term, your support will be remembered.

Remember, a discriminatory climate often forces women to adopt aggressive behavior patterns in order to advance through the ranks. Unfortunately, the aggression is not always channeled correctly into positive change. While there is no one right solution that will enable women colleagues to overcome problems like the ones outlined above, better communication skills can help establish shared goals and overcome common obstacles.

Gripe Sessions Can Be Profitable

When you suspect that there are problems between the men and women on your staff, staging a no-holds-barred discussion may be the answer. For one thing, such a session underlines your concern for everyone's feelings. Framing the discussion in a business context underscores that individuals may have to put their personal prejudices and feelings aside for the good of the entire group.

To get the maximum results from such a meeting, however, follow these suggestions:

• **Keep the meeting small.** The larger the meeting, the easier it is for people who are reluctant to speak up to fade into the crowd. They remain silent while others talk. Yet, they may be just the people whose ideas and problems should be aired.

It's not easy to "hide," though, at a small meeting, so if you have a large group of people reporting to you, have two or three small sessions rather than one all-inclusive one. You'll get a lot more straight talk even if you have to prod.

• **Hold the session "in the round."** If there is a round table available at which the group can be seated comfortably, this is ideal. If not, arrange the chairs in a circular fashion. The aim is to have high eye contact, because this leads to maximum interaction. People are much more likely to respond when they can see one another. Furthermore, there is no "head" of a circle. Thus, you fit into the group as a participant rather than a boss and will not inhibit your subordinates.

• **Listen intently.** This is one situation that you don't want to dominate—it's their talk session, not yours. Still, you don't want to come across as uninvolved or uninterested. So, when someone asks you a question or looks at you expectantly for a response, paraphrase what's been said as you begin your reply.

• **Put it in writing.** From the very beginning, take notes—this will indicate to members of the group that you are indeed serious about hearing what they have to say. It can also be viewed as a hopeful sign that you are going to do something about particular items that they mention.

• **Use a flipchart or blackboard.** This is a technique that can loosen the tongues of the most reluctant participants. If the discussion is slow in getting started, or if it begins to bog down, get up and go to the chart or board. Then lay the groundwork: "Okay, let's get some important points down—just throw out anything that's bothering you, anything at all."

• **Maintain your credibility.** All these factors are important in keeping an "open" session alive and kicking. But what about the next one, will it move, too? The answer to that depends on what you do about what you've heard at this session. So, as it draws to a close, review your notes aloud to make sure that you have everything straight. Then follow up on it and report back to the people concerned. This is the insurance for your credibility—and for another meeting that will be just as profitable as the first.

Romance at Work

Sometimes, that most wonderful of all situations arises on the job—you meet Ms. or Mr. Right, fall head-over-heels, and plan to spend the rest of your lives together. When this happens, the rest of the issues have a way of sorting themselves out—when and how to tell co-workers, and whether one of you should leave the firm in order to preserve healthy business and domestic relationships.

But other sexual interactions pose more questions. When someone makes a sexual overture toward you, you had better answer the following questions:

• **Is it a power play?** Since Biblical times, people have been using sex—either promised or provided—to advance themselves. Look objectively at the person's position and decide whether he or she may need your support for some political end.

Then look at your own emotions: If you feel terribly complimented, empowered, or excited by the sexual advance, take care that your own emotions don't lead you into a trap.

• **Is it just sleazy?** Men and women who are bored and looking for sexual liaisons among co-workers have failed to assess their professional or personal priorities. Avoid them.

• **Is it interesting?** It would be crazy to tell people never to enter into relationships with co-workers. If you have determined (after waiting a reasonable amount of time and doing some reflecting) that the personal and political risks are minimal, you may decide to make an informed decision to explore the attraction—but always do it outside the workplace. However, be aware that there are always more risks than you can anticipate when romance and work are combined.

• **What is the context?** If a co-worker invites you to dinner or the theater, the flirtation has begun on a fairly sophisticated plane and is mutually agreed upon. However, a sudden advance at the office party—allowing no time for you to consider the ramifications—should be rebuffed at all costs. It can easily make you the center of gossip and destroy the credibility it may have taken you years to achieve.

Assess Your Company's Reaction

Before you become involved in a romantic relationship at work, determine whether the powers that be will frown on it, grudgingly accept it, or look the other way. Although companies vary in their outlook, the following information, culled from the responses of numerous corporate executives, both male and female, may be helpful to you.

Do many companies have policies on office romance?

Of 100 female executives interviewed, only 12% said their companies did. But 86% of those whose company cultures were conservative said their management informally discouraged employee dating, while 50% of those at companies where the culture was more open felt that office romance was neither actively encouraged nor discouraged.

If you work in a conservative industry such as banking or insurance, does that necessarily mean romance is taboo?

It depends more on company values. If personal and professional lives don't frequently crisscross—for example, if the company doesn't provide opportunities for people to get together outside the office, and people don't share details about their personal lives with each other—a relationship between two employees probably won't go over well with management.

Does that mean that romance is more likely to bud when people work long hours or take business trips together?

Definitely. The sheer proximity of others and the intensity of work are factors that seem to create natural conditions for attractions to develop. And management is more likely to tolerate it, possibly because it happens more under these circumstances. They realize that trying to legislate it away would be futile.

Are companies that promote socializing generally more open to romances that ensue?

They usually don't discourage them. The exceptions would be romances between a boss and an employee, because of the bad effect they could have on morale if others perceived that their colleague was getting preferential treatment as a result.

Is it too much pillow talk that concerns management?

Sure. Many companies are concerned about information leakage, even within their own company. So if you're thinking of dating someone and either one of you is in the position of knowing sensitive information, be aware that management may step in if there's an appearance of impropriety.

And, needless to say, your job may be in jeopardy if you become involved with a client or a competitor.

How good an indicator is a company's attitude toward past employee romances in regards to how your own will be treated?

If the relationship was recent and your situation is analogous—that is, your job levels and departments are similar—it may be a good indicator.

Are women more likely to get hurt professionally than the men they're romantically involved with?

Again, it depends on company culture. In more conservative firms studied, 68% of the women interviewed felt the risks were greater for women. But at more open companies, half said it applied equally to men and women.

Assess the Risks to Your Career

If you are becoming involved in an office relationship, you should consider the tension it may cause your colleagues and the repercussions it might have on your career:

• **It's difficult to keep romantic involvements a secret.** That doesn't mean you shouldn't be discreet by keeping a professional distance and a "my work comes first" attitude in the office. Although men often want to let others know whom they're dating because it adds to their power or masculine image, it's usually in your best interest to try not to let on.

• **Co-workers may think that you're being manipulative.** If your romantic liaison is with your boss, charges of favoritism that are eventually brought to management's attention may force one of you out.

• **Once others suspect your involvement,** you may be cut off from—or deliberately fed—office information or gossip because it's presumed that you have the boss's ear.

If you begin seeing each other outside working hours, discuss a change in jobs to avoid any possible conflict-of-interest problems.

If your boss approaches management regarding a transfer for you and is honest about the reason for presenting such a request, your boss will probably be respected for showing concern about remaining impartial and effective on the job, and you'll be able to pursue the romance without the potential negative effects of remaining in a boss/subordinate situation.

Your should know, however, that becoming romantically involved with your boss is generally frowned upon. Psychologists who have studied the effects of such liaisons say it doesn't do either party any good.

We surveyed 2,200 people and 10% of our respondents said that they have been romantically involved, and one out of three said that the romance had a positive effect on his or her job. The three big pluses were: making the job more enjoyable, opportunities for career advancement, and contributing to self-growth and development.

One of our respondents told us: "I dated my boss for three years and worked for him the entire time of my employment. For us, work was work and fun was fun and we managed to keep the two things separate. It was much easier for me to express my opinions and ambitions in the office, and that made my job much more pleasant. I was even able to continue working for him after I broke the relationship off."

One-fourth of the respondents who had some romantic involvement with their colleagues, however, reported negative professional consequences. The worst scenario—losing their jobs—happened to just a handful. Two also reported that their partners were fired as well. Beyond that, the most often-mentioned problems were that the romance interfered with productivity or concentration at work, or made it difficult to keep personal and professional lives separate.

One troubled reader reported that: "Most bosses ask their staff to cover for them. While occasional little white lies are no big deal, my boss expected me to lie to clients if he didn't come in because he'd had too much to drink or for some other similar reason, and also to back up the stories he told his wife about the time we'd spent together. I began to resent it and after several years, broke off the relationship and left the job."

Office romances resulted in marriage or living-together situations for 13% of those who got involved. "We got married and now I'm the boss," said one respondent.

In the case of two of our respondents, a romantic relationship with the boss had interesting professional consequences and humorous endings. Said one woman: "I was promoted over him and fired him to protect myself."

A second woman wrote: "I own the company now and I fired him."

Sexual Harassment—A Threat to You and Your Company

Besides creating an awkward environment in the office, a sexual liaison could lead to a sexual harassment suit against your company. According to the courts, sexual harassment occurs when an employer alters an employee's job conditions as a result of the employee's refusal to submit to sexual demands. Corporations are strictly liable for such harassment, because the supervisor who requests sexual favors "by definition acts as the company."

Hostile-environment sexual harassment occurs when an employer's conduct has the purpose or effect of unreasonably interfering with an individual's work performance, or creating an intimidating, hostile, or offensive environment." Corporations aren't responsible in that case because no quid pro quo exists.

Sexual Harassment: Your Duty as a Manager

Managers at all levels should take extra care to keep even the appearance of sexual harassment out of the workplace.

What happens when a lower-level manager you supervise is having relations with someone who reports to him or her? Are you obligated to say something to this manager to protect your company?

The answer is yes and no. You need to say something, but you are well advised to watch how you say it. There are three potential problems:

1. A case of sexual harassment. If the subordinate later charges your assistant with having pressured him or her into sexual activity, your company could be faced with unpleasant publicity and an expensive legal process.

Even if the employee now seems to be giving full assent to the relationship, you can't relax. The subordinate may later claim that he or she was "forced" into an appearance of consent. This employee might not be able to make the charge stick, but throughout the unwelcome publicity, your bosses won't appreciate your having turned a blind eye to the relationship.

2. Favoritism. If your assistant is protective of the subordinate, giving this person the better assignments, or showing favoritism on the job, other employees will believe the situation is unfair to them. And if the subordinate seems to take advantage of the relationship by working less hard, co-workers will resent it.

3. Subversion. Other employees who disapprove may try to isolate the subordinate from the rest of the work group, or subvert your assistant's effectiveness by deliberately not following his or her instructions.

Your statements to your assistant are hampered by his or her discretion. If this manager hasn't actually displayed affection on the work scene, you are operating on hearsay, even though the knowledge of the affair is widespread.

Some recommendations:

Call your assistant into your office as casually as possible. Be friendly with your invitation so as not to feed the rumor mill.

"Carl," you might say, "I think you should know that there are rumors going around about an affair between you and Anne. True or not, it's important that these rumors not get in the way of your relationships with the others.

"Your personal life is none of my business, but what happens here definitely is. For example, I can tell you now that if Anne were to charge you—and through you, us— with sexual harassment, everyone would suffer: you, me, the company.

"Anything that even looks like favoritism between you and Anne is going to be picked up by everyone else. If your employees don't do their best for you, then your record here is going to suffer. And I won't tolerate lowering the standards, for you or anyone else.

"Finally, I would be very unhappy to get calls from suspicious or angry spouses asking whether I know what is going on."

Carl may wish to discuss the affair with you, now that you have opened the door. Our advice is to stick with what you've said, and decline to discuss the liaison.

From this point on, you want to stay alert to any slippage in performance on the part of either, but you don't want to spy. Ordinary monitoring will probably be sufficient. And you won't want to encourage the rumor-mongers. If an employee comes to you with the latest story, simply refuse to listen.

Conclusions

Aside from your personal relationships with family members and friends, the relationships you have with those you work with rank high in importance. That fact is not surprising. You may see your relatives only for a few hours on special holidays. But you see your co-workers eight hours or more nearly every day. You share with them a common sense of purpose that revolves around the goals of your company. Your business colleagues become, in a sense, your extended family. You enjoy being with them, you depend upon them.

When your work relationships are solid and based on mutual respect and understanding, you are more productive. You have the support and cooperation of your staff and can concentrate on getting your job done well. On the other hand, when a work relationship goes sour it can wreak havoc with your professional and personal life. Worrying about a co-worker who sandbags you at every interval or a subordinate who threatens you with a lawsuit takes your mind off your work.

It makes sense, then, to work at establishing and maintaining positive, healthy work relationships as a way not only that you can advance your career but also that you can avoid the stress some executives encounter when work liaisons become strained.

Here is a checklist you may want to refer to as you reassess your working style as well as your working relationships:

1. Reevaluate your appearance. Make sure you dress

as if you mean business.

2. Look at your office. Does it appear to house some-one with authority?

3. Review the rules of office etiquette and protocol. Be sure to take note of other employees' preferences and prejudices.

4. Establish a business network. Resolve to make a certain number of new contacts each month.

5. Don't forget older contacts and friends. Take the time to nourish these relationships.

6. Relax outside the office. Remember the importance of socializing with those on your team.

7. Reappraise your feelings about working with and managing women. Make sure you aren't holding others (and yourself back) by discriminating among your workers.

8. Consciously work on your management style. Try to strike a balance between being fair and tough, relaxed and demanding.

9. Establish ties with those above. Should you have a mentor? Could you improve the relationship with your imme-diate boss?

10. Review your ability to deal with a special situa-tion: an employee's tears, a public insult, staff members who quarrel.

11. Listen to your employees. Consider mechanisms that will allow them to discuss their problems.

12. Watch for romantic relationships among your staff members. Don't assume everything is innocent. Sexual harassment suits are a growing threat to managers and their companies.

- ■ **Observation:** Your working relationships, how well you interact with the men and women in your office environment, may have as much to do with your eventual success as how well you do your job. In fact, the two are interrelated. If you demonstrate that you respect your workers as individuals based on their abilities and not their sex, you will earn their loyalty. You will be able to count on their help during both good and bad times, and that support will aid your advancement immeasurably.